Space and Time

AUDREY E. RANDLES

Introduction

Existence of space declares the existence of time. Speed, acceleration, and changes define time in our dynamic world. The relative difference between our perception of time and space makes our world dynamic for us.

Space and time are complementary and contra-directed.

Space is associated with actuality.

Time is associated with latency and potentiality.

Space is built by energy and associated information representing the forms of energy, such as mass, kinetic energy, and other energy representations acting in the volumes of the existing objects and systems in the period of time now occurring.

Time is built by the potential energy that is coded and fixed by the potential information.

Furthermore, Space-Time may be reversed. Space influences the direction of the time flow.

The theory of Matrix is a cosmological theory considering the new understanding of space and time, along with the Space-Time and energy structure of the existing objects and systems, including the Universe, visually 'empty' spaces, stars and galaxies, subatomic particles and holes, and other objects and systems existing in our dynamic world.

We introduce a new concept of the multimodal Space-Time and offer an opportunity to access Space-Time characteristics of the multimodal objects and systems of the objects existing in the Universe.

We analyse the multimodal structure of the objects and systems and modes of their functioning in different modalities of Space-Time.

We discuss some principles of time travel.

The Theory of Matrix series of books offers the exiting developments in cosmological theory. 'Space and Time' is the second book of the series.

Discover the secrets hidden deep in the Universe.

Stay well, and enjoy your reading.

Yours sincerely,
Audrey Elizabeth Randles
JULY 29, 2020

Image 1: 'Take a Splash Into the Cosmos' Image credit: NASA/JPL-Caltech

Acknowledgement

We would like to express our gratitude to the National Aeronautics and Space Administration (NASA), NASA's Jet Propulsion Laboratory (JPL), the Infrared Processing and Analysis Center (IPAC) and the California Institute of Technology (Caltech) for the impressive space image and exciting descriptions chosen for this book.

The views and opinions of the author, expressed in this book, do not necessarily state or reflect those of NASA's Jet Propulsion Laboratory, the Infrared Processing and Analysis Center, the California Institute of Technology or the National Aeronautics and Space Administration.

Image 2: 'Sun Over Earth's Horizon' Image Credit: NASA

Contents

Introduction
Acknowledgement
Chapter 1   Space and Time
Chapter 2   Pre-existing Concepts
Chapter 3   The Meaning of Time
Chapter 4   The Space-Time Coefficient
Chapter 5   Time Travel
Chapter 6   'o' Space-Time Point
Chapter 7   1-dimensional Space-Time
Chapter 8   Space-Time Relativity
Chapter 9   Information and Energy
Chapter 10   2-dimensional Space-Time
Chapter 11   Space-Time Equivalence
Chapter 12   Space of the Current Time
Chapter 13   Multimodal Space-Time
Chapter 14   Energy Shapes Space-Time
Chapter 15   Space-Time Arrow
Chapter 16   Matrix Balance
Chapter 17   Space-Time Imbalance
Chapter 18   Gravity and Antigravity
Chapter 19   Balancing forces
Chapter 20   The Main Principles
Chapter 21   Prospects
Afterword
Content Use Policy

Space and Time

Space does not exist independent of objects and systems. Space, or a volume, is the specific property of an object or a system of the objects.

Neither empty space nor empty time exists in the Universe. Space and time are built by and filled with energy. The 'outer space' is always another object or a system. Space is associated with actuality. It is built by energy represented in mass, kinetic energy, and other energy representations acting in the volumes of the existing objects and systems in the period of time now occurring. If an object's volume exists now, we can measure it. Still, the volume of the object in the past or its future changes are inaccessible to the direct measurement we undertake at this moment.

Time is the property of the existing objects and systems, similar to other features, such as their volume and weight. Time is built by the potential energy, which is coded and fixed by the potential information setting the form of energy. Time is associated with latency or potentiality. We know about the timeline connecting the past and the future, but we cannot directly perceive and measure the past and future events, and so we dream about time travels.

In our daily life, we apply the point-of-time settings, associated with our perception of time as a point 'now', and the linear-time settings following our understanding of time as the time 'flow' moving steadily from the past to the future.

Space and time are complementary and contra-directed. Space-Time may be represented as a coexistence of space and time, building the space system and the time system in the opposite direction. Space appears to contradict the time claims. Space influences the time flow. Time slows down when a space object approaches the higher speed, such as the speed of light. A clock, measuring time in a moving frame, will be seen to be 'dilated' relative to the 'proper time' in which the clock is at rest. Time will always be shortest, we would say - fastest, as measured in the rest frame in which the space object is at rest.

Space may be transformed into time, and time may be transformed into space. Space-Time transformations are the function of energy. The simple example of these transformations happens when you take your coins and put them in a bank. The volume of the existing coins becomes the potential space that influences your life satisfaction and credit rating. Your coins become the time-related potentiality if the bank cannot process the ordinary commercial transaction you require. The reverse process - the transformation of time into space, supported by the associated Mass-Energy transformations, requires your money, existing in the potentiality of your bank account, become the volume of the hard coins in your hand.

Accordingly, Space-Time does not exist independent of the objects and systems. It's a property of the objects and systems built by the objects. Space-Time is to be defined in relation to a frame of reference. Objects and systems of limited volumes are limited in their time of existence and, accordingly, limited in Space-Time. Objects and systems of limited mass are limited in Space-Time.

Pre-existing Concepts

Pre-existing Space-Time concepts can be found in the scientific publications by Paul Langevin, Hermann Weyl, Hermann Minkowski, and Albert Einstein. To sum, 'any event is determined in terms of its position in space and time by four coordinates measured in this reference system, three for space and one for time.'

Paul Langevin suggested: 'We can define time by all the events that follow one another at one point, for example in the same portion of matter in relation to a reference system, and define space by all the simultaneous events. This definition of space corresponds in effect to that the shape of a moving body is defined by the set of simultaneous positions of the various portions of matter it contains, its various material points, or by all events posed by the simultaneous presence of these different material points.' [Langevin Paul, The Evolution of Space and Time (1911)]

In 1920 Hermann Minkowski wrote: 'Henceforth, space for itself, and time for itself shall completely reduce to a mere shadow, and only some sort of union of the two shall preserve independence.' [Minkowski Hermann, Space and Time (1920)]

Hermann Weyl writes: 'My body is in any instant of my life at a certain world-(= space-time-)location; thus it traverses a one-dimensional succession of world points, a world line, as well as any other body.' [Weyl Hermann, The

Discussion concerning the theory of relativity at the Meeting of Natural Scientists (1920)]

Currently, there are multiple concepts and essays on the nature of time and time-related issues. Time is considered as one of four dimensions of the world - the addition to three space dimensions. It is conceptualised as an absolute simultaneity of events.

There are multiple Space-Time descriptions of causality. Time is replaced by the relation of cause and effect, the temporal order of cause and effect, bidirectional causation, causal asymmetry, causal loops and backward causation. Some theorists deny that time really passes.

We introduce new concepts of Space-Time and offer an opportunity to access the Space-Time structures of the multimodal objects and systems, characteristics of gravity, antigravity, and other aspects associated with Space-Time.

The Theory of Matrix applies to every existing object and system, including visually 'empty' spaces, subatomic particles and holes, and the Universe.

Albert Einstein has specified the connection between mass and energy, '... a body of mass m is to be regarded as a store of energy of magnitude mc^2.' [Einstein Albert, A Brief Outline of the Development of the Theory of Relativity (1921)] Reading the famous equation for mass-energy equivalence $E = mc^2$ with respect to the 2-dimensional Space-Time helps consider the Theory of Matrix.

The Meaning of Time

The human perception of time as a time-point, the current 'now', is the main boundary we operate in. If you look inside your conscious, intelligent operating system, you notice that it is always 'now' for you. We live now, and now we think about past or future events. We exist at the '0' time-point between the past and the future.

Following our understanding of time, we operate on the imaginary timeline of a system, such as our planet. We think about time as the continued existence. We build the lines of our daily time according to the 24-hour cycle of the Earth rotation and four seasons (spring, summer, autumn, and winter) marked by the particular weather patterns and daylight hours, resulting from the orientation of the Earth's tilt with respect to the Sun.

We are using a uniform standard time within time zone established in a region with the orientation on the standard meridian of the Earth and longitudes. The duration of a person's life we count by tradition in years - the time taken by our planet to make one revolution around the Sun. Accordingly, we act on the timeline of the system.

Our conscious 'I' - the heart of the individual core, operates as a 'point', similar to a point in time - a moment. Our momentary perception of time might be compared with the series of dots we perceive on the line made on one side of the 3-dimensional time-cube of the 3-dimensional world.

The human perception of time is reflected in mathematics and other abstract sciences associated with numbers, quantities, and space. Mathematics, based on single numbers, where 1^3 still equals 1, does not overcome the barrier of pointed consciousness. Algebra, being a part of mathematics in which letters and other general symbols are used to represent numbers and quantities in formulae and equations, exhibits similar limitations.

Our momentary perception of time is applied to other disciplines, such as physics. Time characteristics in theoretical physics become incorporated, for the sake of simplicity, into the concepts of simultaneity and cause and effect replacing time dimensions. Some scientists offer the hypothesis of the time non-existence and replace time dimensions with 'point events' reflecting time.

Replacing time dimensions we build the cause and effects lines in our sciences and our lives, and we are still unsatisfied in predictions of future events. Without knowledge of time, we have chaos.

The above confirms the importance of time in our lives and our sciences, and our refusal to see the truth is the sublimation and repression of the time-related psychological conflict in men.

All dynamics of our dynamic world is associated with the asymmetry between our perception of space and time. The relation between 1-dimensional space and 1-dimensional time, we sense as the speed. The disproportion between space and time and, accordingly, Space-Time and energy imbalance within the system, we regard as gravity, and we perceive balancing transformations as acceleration.

We ask ourselves, what makes our world so beautiful, dynamic, changeable, variable, multiplex and complex in its nature and effects?

The normal biologically driven limits to human perception of time draw our dynamic and multiplex world.

We find astonishing the beauty of a snowflake and the elaborate structure of a leaf, which are never duplicated by nature. The infinity is flying up to the infinite number of points. The infinity is dropping down, dividing a point into the infinite number of parts. We are counting space to the eleventh dimension and looking forward to having the infinite number of the space dimensions ahead.

Running before the wind, we can experience the Abyss, Continuum, 1-dimensional infinity, and other exciting things. It is similar to using a sledgehammer to crack a nut. We, humans, have got stuck solving the riddles of the Universe, and we are unable to progress with our tasks unless we see our biological and psychological limitations on time perception. The world is different, and we cannot access it from the 'pointed' point of view. The return to the pointed, linear or flat space or time would be the sign of regression in our sciences.

Sometimes the dreamers think about the Universe as a computer game with humans as the charged particles producing electric current for this computer. We disagree.

The world is real. Energy, matter, forms, time, and space certainly exist. But we can only deal with the very limited modality of the Universe - our Universe as we know and sense it. The complement, or, say, the other multiple modalities of the Grand Universe, are currently inaccessible to humans.

Image 3: 'NEOWISE Opens its Eyes' Image credit: NASA/JPL-Caltech   Instrument: NEOWISE Telescope

'It shows a patch of sky in the constellation Canes Venatici, or the Hunting Dogs. The galaxy seen near the center is NGC 4111, the largest member of a small group of galaxies located 50 million light-years away... NGC 4111's core contains a supermassive black hole actively feeding off surrounding gas and dust.' NASA

The Space-Time Coefficient

The concept of relativity and Einstein's Special Theory of Relativity state that all motion is relative and the velocity of light in a vacuum has a constant value which nothing can exceed. We add that the world dynamics reflects the asymmetry in our perception of space and time.

The speed of light in a vacuum sets the fundamental relation between 1-dimensional space and 1-dimensional time that defines the exact proportion of space to time as the upper limit of our dynamic world. Planck length, Planck time, and Planck energy, based on the calculations using the speed of light in a vacuum as the fundamentals proven by experiments, define the exact lower limits of the existing objects and systems and the Universe as we measure and sense it.

Speed as the rapidity of movement represents the relation between 1-dimensional space and 1-dimensional time. The Space-Time Coefficient is a factor that measures the exact proportion of this relation. The Space-Time Coefficient [c] equals the numerical value of the velocity of light in a vacuum and, accordingly, it is the numerical value of the speed of electromagnetic waves propagation in a vacuum. The value of the speed of light in a vacuum (c) equals 299 792 458 m s-1 (meter per second). It is a CODATA Fundamental Physical Constant. The velocity of light measurements carried out by Michelson and his associates over the period 1924 to 1935.

The Space-Time Coefficient [c] applies to the decisions associated with the transmission of the Space-Time, Mass-Energy, and Info-Energy transformations, including those reflected in gravity and antigravity propagation in our dynamic world. 'Relativity theory ... shares with the corpuscular theory of light the unusual property that light carries inertial mass from the emitting to the absorbing object.' [Einstein Albert, The Development of Our Views on the Composition and Essence of Radiation (1909)]

The speed of gravity propagation was introduced by Hendrik Lorentz (1900) and confirmed by Hermann Minkowski (1908) as follows, 'The law of mass attraction, which has been just described and which is formulated in accordance with the relativity postulate, would signify that gravitation is propagated with the velocity of light.' [Minkowski Hermann, The Fundamental Equations for Electromagnetic Processes in Moving Bodies, Appendix (1908)]

In our dynamic world, the Space-Time, Info-Energy, and Mass-Energy transformations, including those reflected in gravitational acceleration and anti-gravitational deceleration, propagate with the speed of light. Background radiation and light carry inertial mass from the emitting object to the absorbing object.

Dynamic representations of background radiation, including those currently known as cosmic background radiation, such as the cosmic microwave background radiation and infrared background, and light provide a mechanism for the transmission of the Space-Time, Info-Energy, and Mass-Energy transformations in our dynamic world.

Time Travel

We mentioned above that in our daily life we apply point-of-time settings, reflecting our perception of time as a point 'now', and linear-time settings following our understanding of time as the time 'flow' moving steadily from the past to the future. We cannot perceive directly the objects and systems, events and their changes, which lie in the past and the future. We cannot at this point have dinner eaten last night. We cannot now touch the book we will first glance at tomorrow. Accordingly, some people dream about time travels to influence upon the past or observe future events.

The first point of this discussion is related to our perception of space objects and systems. Space objects are associated with the actuality, current 'now'. We use our space-associated personal bodies and space-objects, such as bikes, cars, and aeroplanes, in our journeys in space. We suppose to use our time-associated personal bodies and time-objects, which we cannot currently perceive, for our adventures in time.

We had learned to sense and manage space-associated bodies and, by the same token, we can develop our time perception and learn to detect and manage our time-associated personal bodies before the first conscious time journey. When our consciousness operates in time, the perceived volumes of our time bodies are not much different from their space volumes.

Besides, we are not self-aware enough to operate on our unique timelines. We mentioned above that following our perception of time, we operate on the imaginary timeline of our planet. We count the duration of our lives in years - the number of revolutions made by the Earth around the Sun. Time and time-associated energy and information are currently inaccessible to humans. Objects and systems, primarily developing time, time-related energy and associated data, are currently latent to us. Their latency is related to the limits of human perception of time and time-related Info-Energy (forms of energy).

Our memory helps us reconstructing the past, and our mind can estimate the future, while our senses fail in perceiving events directly at any time distance.

The second point of this discussion is related to the incompatible Space-Time settings people expect to work for time-travels. When people imagine a time-traveller, they move the personage into the past or the future of the system, creating the incompatible Space-Time settings that would not work in time-travel. They imagine a time-traveller acting in a new time environment, sometimes belonging to a former time of the Earth - the Space of the Regressive Time of the system. The Space of the Regressive Time does not have qualities of the Space of the Current Time of the system. It cannot form the space-related bodies of the traveller and the planet.

We would represent energy and force, associated with time and supporting the time travel as follows,

$E(t) = m \div c^2$

$F(t) = E(t) \div t$,

where t is time, E is energy and F is force, c is the speed of light in a vacuum.

The perceptible time travel is possible if the centres of the time-traveller's Matrix and the Matrix of the system he travels, such as our planet, are united. For example, the time-traveller and the planet act at the same point 'here' in space. It makes the time-bodies of the time-traveller and the planet available for interaction. Similarly, we can travel in space, because the space-traveller and the space-body of the planet interact at the same moment 'now' in time.

If the time travel had happened before the ability to perceive time was developed in humans, a traveller would not know that he had moved to the past, the future or any other direction. He would continue acting in the Space of his personal Current Time, his personal 'now'. Accordingly, it would not be the past or the future on the timeline of the system but the period of time now occurring.

I would like to stress the critical point - we all are time-travellers without informed consent, so to say. We cannot stop travelling in time in our dynamic world. The aim is to reach meaningful destinations.

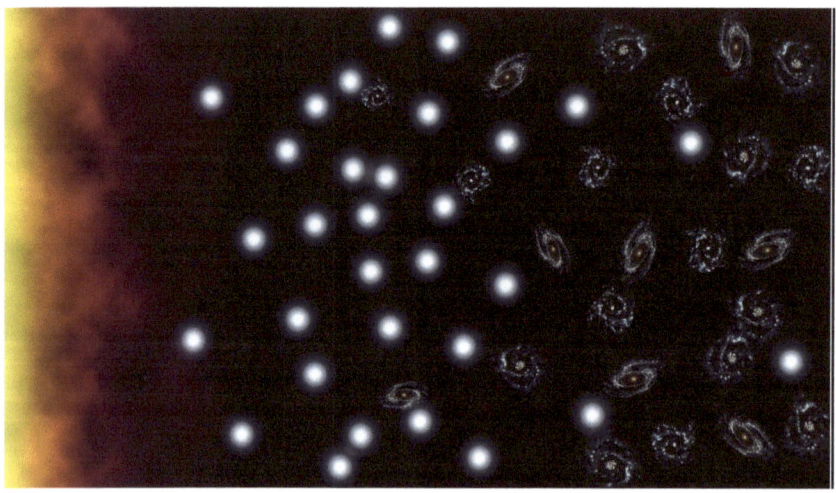

Image 4: 'Baby Galaxies in the Adult Universe' Image credit: NASA/JPL-Caltech

'o' Space-Time Point

We mentioned above that the time-traveller and the planet must act at the same point 'here' in space, making the time-bodies of the time-traveller and the planet available for interaction.

Consciously we operate at the 'o' time-point dividing the past from the future. 'Now' you are thinking about the past two months. 'Now' you are planning your trip to London and 'now' you are reading this book. The 'o' time-point provides us with the volume of an object. We perceive objects and systems as displaying three spatial dimensions. Our conscious 'I', the heart of individuality, operates as a 'point', similar to a point in time - a moment.

The 'o' time-point is a time component of the 'o' Space-Time point.

The 'o' Space-Time point, or 'a space-time-null-point', is a point 'here' and 'now'. The point 'here' and 'now' as a 'o' world point was mentioned in Special Relativity in association with the mathematical model of the Light Cone. 'Let us call any world-point o as a space-time-null-point.' [Minkowski Hermann, Space and Time (1920)] 'A world-point is a 'here-now'. [Weyl Hermann, The Discussion concerning the Theory of Relativity at the Meeting of Natural Scientists (1920)]

The 'o' Space-Time point may be represented as a coexistence of the 'o' space-point and 'o' time-point, or 'here' and 'now'. The 'o' space-point and 'o' time-point are complementary building the time system and the space system in the opposite direction.

Theoretically, a 'o' time-point, or 'a time-null-point', permits maximum space. For example, all the Universe exists at this point - this moment in time. It exists 'now', at the same moment for all the Universe with all its space, masses, stars, and galaxies.

Furthermore, a 'o' space-point, or 'a space-null-point', is a point 'here'. A 'o' space-point permits maximum time. For example, all the time of the Universe exists 'here' - at this point in space. It exists at the end of the needle - the same point in space for all potentiality of time with all its energy and information, with all the past and unlimited opportunities of the current time and the future. We could build the Matrix at a space-null-point - the point 'here' and the associated time volume. This Matrix would represent the description of the unlimited latent dimensions coexisting in the state of a balance.

The coexisting 'o' space-points and 'o' time-points cancel out one another at the 'o' Space-Time point, along with the start and the end of space and time. It opens the unlimited actuality and potentiality, nullified in the ultimate balance of the Grand Universe.

We think about the Universe as the infinite space, existing in various combinations at this moment and every other moment in time. We cannot imagine the Universe as the infinite time that exists in different arrangements at this point and every other point in space. The space of the Universe is filled with a multiplicity of space-objects. The time of the Universe is filled with a multitude of time-objects. The objects and systems, mainly developing time and potential space and accumulating tremendous potential energy resources, are still invisible in the dark.

You remember Hermann Weyl expression, 'My body is in any instant of my life at a certain world-(= Space-Time-)location; thus it traverses a one-dimensional succession of world points, a world line, as well as any other body.' [Weyl Hermann, The Discussion concerning the Theory of Relativity at the Meeting of Natural Scientists (1920)] The coexisting '0' space-points and '0' time-points in the multimodal structures of objects and systems are failing our consciousness.

Planck units, such as Planck length, Planck time, and Planck energy, declare the impossibility of the '0' space-points and '0' time-points in the existing objects and systems. Planck length, Planck time, and Planck energy reflect the minimal Space-Time and Info-Energy imbalance in the objects and systems existing in our dynamic world and accordingly, the minimal conditions of an object or a system's existence.

On the image below,
'This swirling landscape of stars is known as the North America nebula. In visible light, the region resembles North America, but in this new infrared view from NASA's Spitzer Space Telescope, the continent disappears.
Where did the continent go? The reason you don't see it in Spitzer's view has to do, in part, with the fact that infrared light can penetrate dust whereas visible light cannot. Dusty, dark clouds in the visible image become transparent in Spitzer's view. In addition, Spitzer's infrared detectors pick up the glow of dusty cocoons enveloping baby stars.' NASA

Image 5: 'An Extended Stellar Family'   Image credit: NASA/JPL-Caltech

'Clusters of young stars (about one million years old) can be found throughout the image. Slightly older but still very young stars (about 3 to 5 million years) are also liberally scattered across the complex...

'Some areas of this nebula are still very thick with dust and appear dark even in Spitzer's view. For example, the dark "river" in the lower left-center of the image - in the Gulf of Mexico region - are likely to be the youngest stars in the complex (less than a million years old).' NASA

1-dimensional Space-Time

The 1-dimensional Space-Time, supported by energy, is the undifferentiated existence of space, time, and energy underlying our dynamic world. The 1-dimensional Space-Time as the one quality continuing existence is a property of the Universe. The 1-dimensional Space-Time Continuum is immobile and unchanging for human perception. In Metaphysics, the Continuum often presents observers with an Abyss or emptiness. Sometimes observers compare it with an Apeiron of Anaximander, the high level of abstraction resembling the Abyss.

Imperceptibility of the Continuum can be compared with the first impressions of a newborn child, who has similar perceptions of the environment at the start of his life, while his perceptual system is not completed. His vision is not objective, and he does not understand the actual difference between dark and light. This reminiscence is often associated with the pre-morbid infinity and infinite Continuum. Then light has come, and understanding is filling up the environment. The self-conscious 'I' receives first impressions of the light and dark and objects.

The experiences of the Continuum, described in different philosophical and spiritual systems, remind the experiences of the 'Age Regression' in terms of Psychoanalysis. In terms of Coresynthesis, which is the frame of the Theory of Matrix, it is the result of the blocked perception of the multimodal, multidimensional world. Comparing these experiences with the mathematical representation of the Minkowski world,

Absolute world, and modern understanding of the Continuum, we provide a parallel with the signs of the blocked perception of time of the multimodal world we exist in. The perception can be blocked intentionally in meditation, unintentionally in hypnosis and some emotional conditions, and it may be blocked as a result of the evolutionary stage, limiting perception of the multimodal world.

The results of our psychological investigation of the Continuum clearly show the undifferentiated space, time, and energy in ultimate dynamics that underly our dynamic world as the continuing duration.

The meaning of aether as an empty space or a substance with the properties of a medium does not apply to the 1-dimensional Space-Time. There is no space in the traditional meaning of the word, but the continuing duration of the one quality undifferentiated space-time-energy. There is no perceptible progression or emission within the 1-dimensional Space-Time.

The asymmetry between our perception of space and time makes the world dynamic for us. We, humans, differentiate the 1-dimensional Space-Time - the relations between 1-dimensional space and 1-dimensional time we perceive as the speed. Accordingly, the 1-dimensional Space-Time Continuum is neither space nor time but the continuing duration of the one quality undifferentiated ultimate dynamics underlying our dynamic world.

Space-Time Relativity

Space-Time and Info-Energy do not exist independent of the objects and systems. Time, space, mass, energy and associated information are the properties of every existing object and every system of the objects. Space-Time and Info-Energy are to be defined in relation to the frame of reference.

The concept of relativity and Einstein's Special Theory of Relativity state that all motion is relative and the velocity of light in a vacuum has a constant value which nothing can exceed.

We add that the relative difference between our perception of space and time makes this world dynamic for us - relations between space and time we perceive as the speed, or motion, and the disproportion between space and time in a system we perceive as acceleration and deceleration.

We add a further remark. Time is relative to space, changing, and transforming into space, and space is relative to time, changing, and transforming into time.

Time dilation is an example of the changing relations between time and space. Time will always be shortest, or fastest, as measured in the rest frame. Time measured in the frame, in which the clock is at rest, is called the 'proper time'. Theoretically, the 'proper time' means time as measured by a clock following the world line - a one-dimensional succession of world points.

According to the Lorentz transformation, a clock in a moving frame will be seen to be dilated relative to the proper time. This statement was confirmed over the years, for example, by monitoring time dilation in a moving frame using the atomic clock. Time 'dilated' in a moving frame relative to the 'proper time' means the proportion of 1-dimensional space to 1-dimensional time is different in the moving frame relative to the stationary position of the object and relative to the stationary observer.

We can compare the speed of an object moving with speed 1 m s-1 (meter per second) and the speed of light 299 792 458 m s-1 (meter per second) in a vacuum:

1 m ÷ 1 s < 299 792 458 m ÷ 1 s (the speed represented as space relative to time), and

1 s ÷ 1 m > 1 s ÷ 299 792 458 m (the speed represented as time relative to space).

We have a higher velocity in space, along with a lower velocity in time for the speed of light compared to the walking speed.

Time is moving with the slowest speed for a photon of background radiation and light in a vacuum, relative to the stationary observer because the proportion of 1-dimensional space to 1-dimensional time is entirely different for the stationary observer relative to the photon.

Time will always be 'dilated', and the length of the space object shortened in the direction of motion as its speed increases - the proportion of 1-dimensional space to 1-dimensional time is changed relative to the stationary position of the object and relative to the stationary observer.

Accordingly, time will always be 'dilated' and the length of a system shortened in our dynamic world, relative to the stationary position of the system following the world line and relative to the stationary observer measuring time with the atomic clock following the world line. The proportion of 1-

dimensional space to 1-dimensional time is changed if the speed is changed. Once we could have a good clock for measuring time changes at the relatively low speed we move in space, we would notice time dilation in our daily life. The object's higher speed in space always means a lower speed in time, or, say, time and space have the opposite directions.

The speed of light defines the dynamic body of the multimodal Universe in our dynamic world. It reflects the fundamental relation between 1-dimensional space and 1-dimensional time. The speed of light and background radiation in a vacuum defines the exact proportion of space to time as the upper limit of our dynamic world. At the same time, the speed of light defines the exact lower limits for an object or a system existence in our dynamic world - Planck length, Planck time, and Planck energy calculations are based on the speed of light in a vacuum.

The theoretical '0' space-point permits a maximum time for an object or a system. Applying this principle to the 1-dimensional Space-Time, we can observe that a portion of space, transformed into the portion of time, increases the representation of time in the 1-dimensional Space-Time in the rest frame in which both the space object and the atomic clock are at rest. The theoretical '0' time-point permits a maximum space for an object or a system. Applying this principle to the 1-dimensional Space-Time, we can observe that time is always be 'dilated' (and the length of the object shortens) as the speed increases relative to the object in the rest frame or the stationary observer with the atomic clock.

It looks like a portion of time, transformed into the portion of space, increases the representation of space in the 1-dimensional Space-Time in a moving frame.

Some concepts theoretically predict that the rate of time reaches zero as one approaches the speed of light. We argue that the theoretical concepts suppose to take into account the meaning of the word 'speed'. The speed equals the distance travelled by the object divided by the duration of the interval. If the divisor equals zero - the result is meaningless. The division by zero is not defined, or say, no speed exists at '0' time.

Space influences the time 'flow' - the faster the relative velocity, the greater the time dilation. Nevertheless, time never reaches zero in our dynamic world. Planck time reflects the factors limiting the rate of time as one approaches the speed of light.

The time, space, mass, energy, and information of the existing objects and systems are relative to the observer. You remember the Schrödinger's cat. Only the observer could say the cat in the box is either alive or dead. Alternatively, the cat is not in the box in the first place.

The observer observes an event and measures it, but the normal limits of human perception of the event and the environment restrict his ability to observe the existing object or the system. The observer's observation abilities are limited in quantity (5) and quality of senses. We, humans, observe the world and the existing objects and systems using our senses, bodies, and memories. We use the receptors of our bodies with their limited ability to respond to the external stimuli, and it takes time to transmit the signal by the series of synapses, compare it with the preserved data in several layers of memories, and produce the sensible response. We reproduce our limited diapason of senses in new technologies extending the quality of our perceptual abilities. Nevertheless, new technologies reflect our limited quantity of senses.

Accordingly, the world is relative to the observer and not defined.

Image 6: 'A Hard X-ray Look at M51' Image credit: NASA/JPL-Caltech, IPAC

'Bright green sources of high-energy X-ray light captured by NASA's NuSTAR mission are overlaid on an optical-light image of the Whirlpool galaxy (the spiral in the center of the image) and its companion galaxy, M51b (the bright greenish-white spot above the Whirlpool), taken by the Sloan Digital Sky Survey. The bright green spots at the center of the Whirlpool and M51b are created by material surrounding supermassive black holes; additional X-ray sources in the vicinity contribute to the emission. The known ultraluminous neutron star is located on the left side of the Whirlpool.' NASA

Information and Energy

Some words about the terminology. Energy is a quality of information, and information is a quality of energy. The information represents a form of energy 'in formation', in dynamics. Information is always associated, processed, stored, or transmitted along with energy or matter. Energy exists in a form set by information, and information has a form fixed by energy. We use the term 'Info-Energy' (forms of energy) to describe the energy and information, which are complementary and cannot be separated. The density of information is proportional to the density of energy associated with the information.

Mass-Energy is 'the mass of a body regarded as energy, according to the laws of relativity' (Oxford Dictionary). We, respecting the laws of relativity, use the term 'Mass-Energy' as interchangeable with the term 'Info-Energy' of an object or a system.

Energy and information are the fundamental characteristics of the Universe and the existing objects and systems. Info-Energy is to be defined in relation to a frame of reference.

Nothing is coming from nothing and going to nothing. Energy and information cannot be lost. They exist in the latent, or potential, and actual forms.

Space and mass are associated with the actuality. You can directly perceive the volume and weight of the book you keep in your hand.

Time and time-related Info-Energy are latent to us. We cannot directly perceive time-related potential energy and associated information. Time and tide wait for no man. Some objects and systems, events and their changes lie in the past. For example, we cannot right now taste the dinner that we had the other night. On the other hand, we cannot, at this point, read a book we will first glance at tomorrow. Both the dinner and the book are latent to us - we cannot touch or see them directly at the moment.

Time and time-related Info-Energy are associated with potentiality as a capacity to perform work.

When the object or the system's capacity to perform work is being realized - energy is associated with space. Similarly, water, filling up an empty container, takes shape. Energy, existing in space, takes perceivable forms creating sensible information. We sense different forms of energy as masses, kinetic energy, and other energy representations acting in our dynamic world in the period of time now occurring.

The total Info-Energy of an object (or a system) is the total energy and associated information, which the existing object possesses in space and time, including its past, present, and future.

2-dimensional Space-Time

General tendencies, existing within the system's 2-dimensional Space-Time, reflect space diverged from time, the past from the future, and the latency of the system from the system's actuality.

The 2-dimensional Space-Time structure of a system is formed by the system's latent, or potential, Info-Energy.

The system's latent Info-Energy forms an infinitely thin filament of the grid-like 2-dimensional arrangement of the system's latent Info-Energy structure. Energy, associated with the 2-dimensional Space-Time, and related information form the 2-dimensional 'container' for the system.

This imperceptible, grid-like 2-dimensional structure is a complete set of information for the associated multimodal system. The system's 2-dimensional Info-Energy grid forms the system's Spaces of Time.

The time-associated latent 2-dimensional Info-Energy of a system draws the fundamental 2-dimensional time associated with the past, present, and future of the system.

The space-associated latent 2-dimensional Info-Energy of a system forms the fundamental 2-dimensional space in the period of time now occurring - the potential space of the system.

The system's potential space is inflated and undetectable. It underlies the volume of the system in the period of time now occurring. The potential space surrounds the system's volume and acts as a boundary, keeping and transforming energy and associated information in compliance with the stored data. The functioning of the 2-dimensional Space-Time and Info-Energy structure of the potential space is similar to the restricted genetic code expression. It is a transmitter between the system's time and time-associated latent Info-Energy, and the system's volume and space-associated actual Info-Energy acting in the volume of the system in the period of time now occurring. The 2-dimensional toroidal systems, mainly developing their potential space and accumulating great potential resources of energy and data, are still inaccessible within the limits of human perception. We can detect the centres of these systems as the supermassive black holes.

Image 7: 'Flaring Black Hole (Artist's Concept)' Image credit: NASA

'This artist's concept illustrates what the flaring black hole called GX 339-4 might look like.' NASA

Space-Time Equivalence

We mentioned above that the general tendencies, existing within the 2-dimensional Space-Time of an object or a system, reflect space diverged from time. Now we can specify the relation between space and time in more details.

Hermann Minkowski described the relation between mass, energy, and time as follows, 'The kinetic energy of a material point is the product of its mass into the gain of the time over its proper-time.' [Minkowski Hermann, The Fundamental Equations for Electromagnetic Processes in Moving Bodies. Appendix (1908)]

Albert Einstein has also specified the relation between mass and energy. 'If an amount of energy E be given to a body, the inertial mass of the body increases by an amount E/c^2, where c is the velocity of light in vacuo. On the other hand, a body of mass m is to be regarded as a store of energy of magnitude mc^2.' [Einstein Albert, A Brief Outline of the Development of the Theory of Relativity (1921)]

We reaffirm the principle of energy conservation. The total mass of a system may change, although the invariant mass - the total energy of the system remains constant.

The universal proportionality exists between corresponding amounts of energy and mass. The universal proportionality factor between corresponding amounts of energy and mass equals the speed of light squared.

We develop the Theory of Light, uncover unusual properties of light, specify the relation between space and time of a system, and introduce Space-Time equivalence.

Please consider the equation for Mass-Energy equivalence $E=mc^2$ with reference to the 2-dimensional Space-Time understanding.

Energy is associated with the 2-dimensional time the same way as mass, corrected with the squared numerical value of the speed of electromagnetic waves propagation in a vacuum, is associated with the 2-dimensional space.

The universal proportionality exists between corresponding amounts of the fundamental 2-dimensional time and the fundamental 2-dimensional space. The universal proportionality factor between corresponding amounts of time and space equals the squared numerical value of the speed of electromagnetic waves propagation in a vacuum.

$Et^2 = [c]^2 ml^2$,

where E is energy, m is mass, l is length, l^2 is the 2-dimensional space, t is time, t^2 is the 2-dimensional time, and $[c]^2$ is the Coefficient of Transformation. The Coefficient of Transformation $[c]^2$ equals the squared numerical value of the speed of light in a vacuum.

The universal proportionality factor, taken as the squared numerical value of the speed of electromagnetic waves propagation in a vacuum, defies the relation between the fundamental 2-dimensional space and fundamental 2-dimensional time of the existing systems.

Rest mass and rest energy remain proportional to one another the same way as the 2-dimensional space and 2-dimensional time remain proportional to one another.

In sum, the universal proportionality exists between corresponding amounts of the fundamental 2-dimensional time and time-associated latent Info-Energy on the one hand and the fundamental 2-dimensional space and space-associated latent Info-Energy on the other. The Coefficient of Transformation $[c]^2$ defies the relation between the fundamental 2-dimensional space and the fundamental 2-dimensional time of the Universe and every existing object and system.

We state the principle of Space-Time transformations. Some amount of time might be transformed into the proportional amount of space, and some amount of space may be transformed into the proportional amount of time. The Space-Time transformation is a function of energy.

We state the principle of Info-Energy transformations. Some amount of potential energy might be transformed into the proportional amount of actual energy represented in mass, kinetic energy and other energy representations acting in the volume of the system in the period of time now occurring. Some amount of actual energy may be transformed into the proportional amount of potential energy. Transformations of information proceed together with the energy transformations.

We reaffirm the principle of Mass-Energy transformations. Some amount of mass might be transformed into the proportional amount of energy, and some amount of energy may be transformed into the proportional amount of mass.

Space-Time, Info-Energy, and Mass-Energy transformations are enacted by the influence of the Matrix forces.

We state the principle of Space-Time conservation. Space-Time of a system is conserved in the system's multimodal Space-Time and Info-Energy structure. The volume of the system may change, although the total Space-Time of the system remains constant.

Space of the Current Time

Describing the Space of the Current Time, we start with the following quotation from famous German mathematician, theoretical physicist and philosopher Hermann Weyl,

'Between active future and passive past, an empty world-area lies, with which I'm neither actively nor passively connected in the instant 0 - The stage of reality is not a stationary three-dimensional space in which things are engaged in temporal progression, but a four-dimensional world in which time and space are inseparably connected. This objective world does not happen, but it exists; a four-dimensional continuum, but neither space nor time.

'Only in the view of the consciousness that crawls upon within the world-lines of the bodies, a section of this world 'lives up' and passes by as an image that undergoes spatial and temporal changes.' [Einstein Albert, Lenard Philipp, Weyl Hermann, Gehrcke Ernst, The Bad Nauheim Debate. The Discussion concerning the theory of relativity at the Meeting of Natural Scientists. (1920)]

When the capacity to perform work is being realized, energy is associated with space. Energy, existing in space, takes perceptible forms which we can sense as masses, kinetic energy, and other forms of energy acting in the volumes of objects and systems in the micro and macro world. The system's actual Info-Energy fills the volume of a system. A volume is the amount of space that a substance or a system occupies, or that is enclosed within a container

(Oxford Dictionary). Similarly, tea, filling up a cup, occupies a volume.

The Space of the Current Time is the actual space of our dynamic world. We cannot perceive directly the volumes of objects and systems, which have existed many years ago and do not exist right now. We can sense directly the volumes of the objects and systems existing now, and we can measure them.

The space of the past, such as the volume of the dinner that we had a week ago, and the space of the future, such as the volume of a book we will buy next year, are not directly accessible to our senses at present - we cannot touch, see or measure them directly.

Our perception of the objects and systems' characteristics, motion, acceleration, forces actually acting in our dynamic world in the period of time now occurring, applies to the Space of the Current Time at the Matrix centre. We sense them directly now, at the '0' time-point. We can measure them in the standard units associated with the current understanding of space, time, mass, energy, and force. Different measurement systems must be applied to the currently inaccessible Spaces of the Progressive and Regressive Time and the fundamental 2-dimensional Space-Time.

Actual space, represented in a volume, is a specific property of every object and every system existing in our dynamic world. The system's Space of the Current Time equals the volume of the system now. The system's potential space, formed by the Matrix grid, limits the Space of the Current Time at the '0' time-point at the Matrix centre.

Multimodal Space-Time

We introduce a new understanding of the world as the multimodal world with reference to the multimodal, multidimensional Space-Time and Info-Energy structures of the Universe and the existing objects and systems.

Every existing single object and every system of the objects, including visually 'empty' spaces, subatomic particles and holes, stars and planets, galaxies and star clusters, systems holding black holes, other existing objects and systems, and the thermodynamic Universe, are the multimodal objects and systems. The multimodal objects and systems carry the complex properties of multiple Space-Time modalities. Characteristics of the objects and systems are various in different modalities of Space-Time.

The 1-dimensional, 2-dimensional, and 3-dimensional structures of an object or a system are integrated into the object or the system's multimodal Space-Time and Info-Energy structure that we call 'the Matrix' for short.

A typical Matrix represents the system with the four main characteristics, such as mass, energy, volume, and time of existence, reflected in the Matrix characteristics. The total Space-Time and Info-Energy are the properties of a system. Accordingly, the Matrixes of the systems contain their total Space-Time and Info-Energy.

Objects and systems of limited mass, energy, and associated information are limited in space and time of existence. Accordingly, the Matrixes of the existing objects and systems of limited mass, energy, and related information are limited in space and time.

A Matrix does not exist without a system. It is the multimodal structure of a system, and, nevertheless, the Matrix is the self-organising system.

The system's multimodal Space-Time and Info-Energy structures are generated as an outcome of the binary operation of two Matrixes or by any of the following - reflection, self-duplication, binary fission, self-multiplication, and incorporation while maintaining anisotropy.

A Matrix is formed by a regular repeated with the equal distances and typically rectangular grid-like 2-dimensional arrangement of the object or the system's latent Info-Energy. The 2-dimensional grid, shaping the system's multimodal Matrix, limits the volume of the system at the Matrix centre. The '0' time-point at the centre of the Matrix reflects the human perception of time as a point 'now' and provides us with an opportunity to perceive the volumes of the objects and systems existing in our dynamic world.

Space-Time is represented in the system's multimodal Space-Time and Info-Energy structure as the Spaces of Time. The meaning of time applies to the Matrix settings as the past, future, and current time of the object or the system existing 'now', in the space of the period of time now occurring.

The Space of the Current Time is enclosed with the 2-dimensional Info-Energy grid at the Matrix centre. It equals the volume of the related system in the period of time now occurring. The 2-dimensional grid connects the Space of the Current Time with the Spaces of the Progressive and Regressive Time.

The Space of the Progressive Time is related to the future of the system. The Space of the Regressive Time is associated with the past of the system. The total Space-Time of the Matrix equals the total Space-Time, which the system possesses for the total period of its existence, including its past, present, and future.

The system's Space of the Current Time is filled with the actual Info-Energy. We can directly perceive and measure the objects and systems' volumes and the forms of actual energy, such as masses, kinetic energy, and other forms of energy acting in the volumes of objects and systems in our dynamic world in the period of time now occurring. The complete actual information is coded and fixed by the system's actual energy and represented in forms of energy and matter in the volume of the system. The density of information is proportional to the density of energy and matter associated with the information.

According to Albert Einstein, 'the ponderable masses will be the determining factor in producing the field, or, according to the fundamental result of the Special Theory of Relativity, the energy density...' [Albert Einstein, A Brief Outline of the Development of the Theory of Relativity (1921)]

Under the Theory of Matrix, the Matrix is not the field but the structure of the system's energy forming its space and time. The ponderable masses are the determining factors in producing the energy density, information density, and multimodal Space-Time and Info-Energy structures of the existing objects and systems.

The space-associated latent 2-dimensional Info-Energy of the system's grid forms the potential space of the period of time now occurring. The system's potential space underlies and limits the system's volume within the multimodal Space-Time and Info-Energy structure of the system.

The time-associated 2-dimensional Info-Energy of the system's grid forms the time-limits of the system's existence. It forms the Spaces of Time. Energy, building time, is latent Info-Energy associated with the past, present, and future of the system.

Describing the Spaces of the Progressive and Regressive Time, we continue the quotation from Prof. Hermann Weyl,

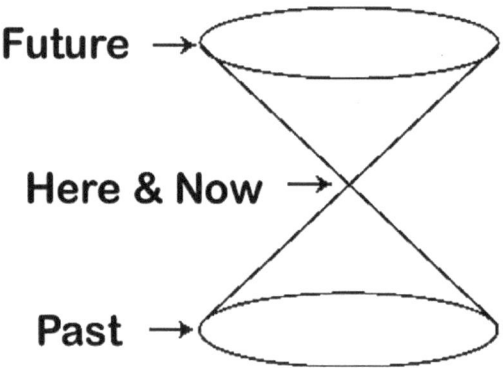

Figure 1: The Light Cone of Special Relativity

'In the front cone all those world points are located that have an influence on my actions in 0, out of it are all the events, that lie finished behind, where "nothing can be changed": the mantle of the front cone separates my active future from my active past.

'The border is formed by the fastest possible propagation of action at all: that of light.

'However, all those events are located in the back cone, for which I have either lively experiences or by which I was informed somehow; only these events maybe have influenced me until now.

'In the exterior of it, however, all of that lies which I will experience or would experience, if my live lasts infinitely or my view might penetrate everything; the mantle of the back-cone separates my passive past from my passive future.' [Einstein Albert, Lenard Philipp, Weyl Hermann, Gehrcke Ernst The Bad Nauheim Debate. The Discussion concerning the theory of relativity at the Meeting of Natural Scientists. (1920)]

There is no priority of the front cone holding the future events nor the back cone holding the past events in the description of the Light Cone of the Special Relativity.

There is neither priority of the Progressive nor Regressive Time in the Matrix of a system. They are symmetric at '0' Space-Time point 'here' and 'now'. The Spaces of the Progressive and Regressive Time look like a hologram. They connect to the undifferentiated Continuum represented as a black background on the images of the Matrixes (Figures 2-9). On the image of the Light Cone, or the Matrix of Light (Figure 1), the Continuum is represented as a white background.

The space-time-energy Continuum fills up spaces of our past and the future. The 1-dimensional space-time-energy is a property of every existing object and every system.

We cannot directly perceive the energy of the 1-dimensional Space-Time filling the Spaces of the Progressive and Regressive Time. They look immobile, unchanging, and inaccessible, appear empty and connected with the undifferentiated Continuum.

The 1-dimensional Info-Energy of duration is psychologically associated with the object or the system's length of existence.

The 1-dimensional Space-Time is not the timeline connecting the past with the future. The cause-effect line, connecting the past event(s) with the event(s) of the current moment and the event(s) of the future, is an illusion related to the human perception of time.

The past is the multidimensional time-dominated Space-Time structure holding the associated latent events. Some of these events have had their being, and other events have been left as the potentiality. The same is true for the potential events of the current time and the future, holding multiple variants of the potential events. For example, you choose between some particular lines of action 'now' and act. By your action, you associate and activate some potential events and manifest them into your actuality. The non-manifested multiple variants of the events and their associations are left in the potentiality.

Let's take a simple example. You have made a choice, and you are having a cup of tea now. By this action, you activated your line associated with the cup of tea - the type of tea, the cup presented to you by your best friend... your birthday... the reminiscence of having a cup of tea in the House of Commons last summer... the idea to visit London... the feeling of satisfaction you will have later on. All other variants of your action, such as getting ready for work, reading the book or sending the email to your loved one - the non-manifested multiple variants of the events, which could be performed instead of enjoying a cup of tea, are left in the potentiality.

Image 8: 'NASA's Galaxy Mission Celebrates Sixth Anniversary' Image credit: NASA/JPL-Caltech

'M33, one of our closest galactic neighbours, is about 2.9 million light-years away in the constellation Triangulum, part of what's known as our Local Group of galaxies...

'This combined image shows in amazing detail the beautiful and complicated interlacing of the heated dust and young stars. In some regions of M33, dust gathers where there is very little far-ultraviolet light, suggesting that the young stars are obscured or that stars farther away are heating the dust. In some of the outer regions of the galaxy, just the opposite is true: There are plenty of young stars and very little dust.' NASA

Energy Shapes Space-Time

Every existing single object and every system of the objects, including visually 'empty' spaces, subatomic particles and holes, systems holding black holes, other objects and systems and our Universe, are the multimodal objects and systems.

The multimodal systems carry the properties of different Space-Time modalities, reflected in the structure of their Matrixes involving a coexistence of different dimensions and different modalities of Space-Time.

The characteristics of the system are different in different modalities of Space-Time.

According to the theory of General Relativity, energy curves Space-Time. Please see Einstein Albert, 'Relativity: The Special and General Theory' (1916). We support this idea and provide descriptions of the main mechanisms as precise as possible for a person functioning in the dynamic 3-dimensional world.

The 2-dimensional arrangement of the info-energetic net-structure, the Matrix grid, is the infinitely thin filament forming the Matrix of every existing object and every system. The 2-dimensional Info-Energy grid contains the system's potential Info-Energy. The total Info-Energy of the Matrix grid is the total potential energy and associated potential information, which the existing system possesses in space and time, including its past, present, and future. Background radiation, including those known as cosmic background radiation, and light arrange the potential Info-Energy of the Matrix grid of the existing objects and systems.

The grid composes the fundamental 2-dimensional Space-Time of the system, limits the system's volume at the centre of the Matrix and shapes the system's Spaces of Time. The potential space of the Current Time forms the 2-dimensional 'container' for the system's volume filled with actual Info-Energy.

Strictly speaking, the reproduction of the Matrix 2-dimensional grid on the paper is not exact. It is related to the difficulties to reproduce the 2-dimensional structure on the paper.

It is a theoretical mistake that one can currently reproduce a 2-dimensional structure on the paper. A drawing on the paper and PC screen does not reflect a 2-dimensional structure - our world-perception influences on making drawings and PC programming. The relative difference between our perception of space and time makes this world of perceptible forms being sensible to us. Accordingly, the structures, which are reproduced on the paper or PC screen, regardless of how special they would be, are no more than our projections of the 3-dimensional objects. Their proportions do not reflect characteristics of the 2-dimensional structures. Similarly, proportions and directions of the 2-dimensional structure in the multimodal Space-Time are entirely different from our projection of any structure on the surface.

Besides, one cannot create a 2-dimensional structure using 3-dimensional particles, one layer of 3-dimensional cells, crystals or atoms, even with the help of nanotechnology. The closest object to the understanding of the 2-dimensional space structure is an ideal square wave.

The exact reproduction of the Matrixes (Figures 2-9) without damaging the main idea is impossible on the paper or in the computer simulation. Although the shapes of the identified types of Matrixes are correct, we rely entirely on the description.

Two types of the Matrixes have been identified - the Time-Rising Matrix (TRM) and the Space-Rising Matrix (SRM). The Matrix grid has a form of the Riemannian Manifold with the positive or negative curvature, and Riemannian geometry applies to its investigation.

The TRMs are the property of the radiating objects and systems, including 'black body' radiation spectrum. The 2-dimensional grid of the TRM has a form of the Riemannian Manifold with the negative curvature (Figure 2). SRMs are the property of non-radiating objects and systems. The 2-dimensional grid of the SRM has a form of the Riemannian Manifold with the positive curvature (Figure 3).

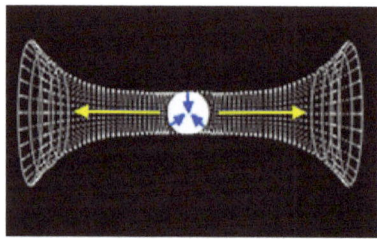

Figure 2: Time-Rising Matrix

We mentioned above that the Space of the Progressive Time is the space of the Matrix cone related to the future of the system (Figures 2-9 - the left cone). The Space of the Regressive Time is the space of the Matrix cone associated with the past of the system (Figures 2-9 - the right cone).

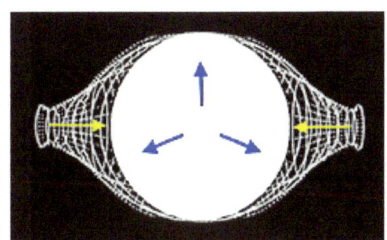

Figure 3: Space-Rising Matrix

The system's grid is the 2-dimensional framework of the multimodal system. It holds a complete set of data related to the embedded system, similar to a chromosome of a living cell. It carries information in the form of Info-Energy blocks.

The notion of the curvature is to be defined in a way that is intrinsic to the manifold. The curvature of the 2-dimensional grid depends on the Space-Time, Info-Energy, and Mass-Energy properties of the related system. The curvature of the Matrix does not depend on how the surface is enclosed in 1-dimensional, 3-dimensional or higher-dimensional Space-Time. The curvature of the Matrix does not depend on how the object or the system's volume is inserted at the Matrix centre.

Black pixels, reading zero and corresponding to the latent information, are fixed by the latent energy of the 2-dimensional grid. They are associated with the timing mechanisms and sweep rates, and the address of a pixel corresponds to its Space-Time coordinates.

We suppose that the blocks of the data, involved in the actuality or located relatively close to the point 'now' in time, are situated closer to the centre of the Matrix, than other latent information. Hypothetically, if the Info-Energy blocks are associated with the Matrix Space of the Current Time, then they might have a 3-dimensional structure. The distances along the paths and angles are to be measured as the characteristics of the system's latent, or potential, Info-Energy.

The 2-dimensional grid is arranged by the 2-dimensional representations of background radiation, including those currently known as cosmic background radiation.

Background radiation, arranging the potential Info-Energy into the 2-dimensional framework within the multimodal structures of the objects and systems, builds simultaneously the internal framework and the dynamic Space-Time and Info-Energy skeleton for every object and system existing in our dynamic world. Background radiation and light constitute the background structure of our dynamic world.

We mentioned above that the object or the system's multimodal Space-Time and Info-Energy structure we call the Matrix for short. It seems reasonable measuring the properties of the Matrix centre and the current characteristics of the system in the standard units related to the current understanding of space, time, energy, and force. Different measurement systems must be applied to the currently inaccessible Spaces of the Progressive and Regressive Time and the 2-dimensional Info-Energy grid that forms them.

Space-Time Arrow

The Space-Time axis is the only axis of the 1-dimensional Space-Time underlying the multimodal Space-Time and Info-Energy structures of the existing objects and systems.

According to the Theory of Special Relativity, the space axis of the Light Cone builds a perpendicular to the time axis. In compliance with our investigations and according to the Theory of Matrix, the Light Cone is an example of the Matrix for a flash of light. The Space-Time axis (other terms - Matrix axis, space axis, time axis) is the only axis of the multimodal Space-Time and Info-Energy structure. The Space-Time axis is the only axis of the Universe.

The best way to present the Space-Time axis is by describing the Space-Time Arrow. The Space-Time Arrow is directed along the Space-Time axis. It reflects Space-Time and energy 'flow' within the 1-dimensional Space-Time.

The Space-Time Arrow displays a coexistence of four tendencies - the tendency towards time and the contra-directed tendency towards space, along with the tendency towards the past and the contra-directed tendency towards the future.

The contra-directed Arrows of Space represent the space component of the Space-Time Arrow and depict the tendency towards space. The Arrows of Space indicate the direction of space and space-associated energy 'flow' in the multimodal Space-Time and Info-Energy structure of a system.

The contra-directed Arrows of Time represent the time component of the Space-Time Arrow and depict the tendency towards time. The time component of the Space-Time Arrow may be specified as the Arrow of the Progressive Time associated with the future of a system and the contra-directed Arrow of the Regressive Time associated with the system's past. Accordingly, the Arrows of Time indicate the direction of time and time-associated energy 'flow' in the multimodal Space-Time and Info-Energy structure of a system.

Presenting the coexisting contra-directional tendencies of the Space-Time Arrow within the 1-dimensional Space-Time, we apply the direct order (Arrows of Time ← '0' → and Arrows of Space → '0' ←) providing an opportunity to observe directions of Space-Time and Info-Energy transformations within high energy massive radiating objects and systems tangled by gravity. We mentioned above that a '0' time-point permits maximum space and, accordingly, a '0' space-point permits maximum time.

Please follow the development of the contra-directional tendencies of the Space-Time Arrow.

The time component of the Space-Time Arrow starts at the '0' Space-Time point at the centre of the system's Matrix as the Arrow of the Regressive Time. It follows the Matrix axis to the furthest point of the Matrix Space of the Regressive Time - the system's past, to become the space component of the Space-Time Arrow.

Simultaneously, the contra-directed space component of the Space-Time Arrow follows the Space-Time axis from the furthest point of the Space of the Regressive Time, reaches the '0' Space-Time point at the Matrix centre to become the Arrow of the Progressive Time.

The Arrow of the Progressive Time starts at the 'o' Space-Time point at the Matrix centre and follows the Matrix axis to the furthest point of the Matrix Space of the Progressive Time - the system's future, to become the space component of the Space-Time Arrow.

Simultaneously, the contra-directed space component of the Space-Time Arrow follows the Space-Time axis from the furthest point of the Space of the Progressive Time, reaches the 'o' Space-Time point at the Matrix centre to become the Arrow of the Regressive Time.

The Space-Time Arrow is described in dynamics for the illustrative purpose only. A mirror effect is possible.

Please apply the reversed order (Arrows of Time → o ← and Arrows of Space ← o →) to observe the coexisting contra-directional tendencies of the Space-Time Arrow and directions of the Space-Time and Info-Energy 'flow' within the non-radiating objects and systems, such as the systems holding transmitting black holes, the Universe, and other non-radiating objects and systems developing peripheral antigravity.

The Arrows of Time and Arrows of Space indicate the directions of the balancing Space-Time and Info-Energy transformations within the objects and systems existing in our dynamic world.

The Arrows of Space

The Arrows of Space are directed along the Space-Time axis, contra-directed, and symmetric at the 'o' space-point - the point 'here', at the centre of the balanced multimodal Space-Time, Mass-Energy, and Info-Energy structure of the object or the system. The 'o' space-point 'here', coexisting with the time-volume of the related object or the system, is currently inaccessible within normal limits to human perception. We perceive space as the volumes of the objects and systems existing at the point of time 'now'. We might remember, for example, that the volume of the particular object has existed 5 minutes before the current moment. However, we cannot directly perceive it. We perceive the world at the point between the past and the future.

The directions of the Arrows of Space are represented by the blue arrows on the reproduction of the Matrixes. They situated following the human perception of space as the volumes of the existing objects and systems. The Arrows of Space reflect the directions of the Space of the Current Time of the objects and systems as displaying three spatial dimensions x, y, and z or a combination of three directions.

The Arrows of Space display some specific characteristics associated with the type of the multimodal Space-Time and Info-Energy structure of an object or a system. The Arrows of Space are directed from the surface of a high energy massive radiating system to the system's centre (Figure 2). The Arrows of Space are directed from the centre of a non-radiating system to its surface (Figure 3).

In the General Theory of Relativity, Albert Einstein described gravity as a consequence of the curvature of Space-Time caused by the uneven distribution of mass. In astrophysics, an event horizon is a boundary beyond which events cannot affect an observer. In our dynamic world, we operate in the Space of the Current Time. As we sum up the ideas behind these statements, we can introduce the boundary of the curvature of the system's Space-Time as an event horizon that surrounds the total volume of the system that equals the system's Space of the Current Time. As such, the external boundary of the curvature of the Earth Space-

Time, affected by the Earth gravity, is the external boundary of the Earth Space of the Current Time. In case of a high energy massive radiating system, such as the Earth, the system's Arrows of Space start at the external boundary of the curvature of the system's Space-Time affected by gravity. We refer the boundary of the curvature of the system's Space of the Current Time as 'the surface of the system'.

The Arrows of Time

The Arrows of Time are drawn as the yellow arrows on the reproductions of the Matrixes (Figures 2-9).

The Arrow of the Progressive Time is associated with the Matrix Space of the Progressive Time and the future of the related object or the system. The Arrow of the Regressive Time is associated with the Matrix Space of the Regressive Time and the past of the related object or the system. The Arrows of Time reflect Space-Time and Info-Energy dynamics within the Matrix Spaces of the Progressive and Regressive Time.

The Arrows of Time are directed along the Matrix axis, contra-directed and symmetric at the '0' time-point at the centre of the balanced multimodal structure of the object or the system. The '0' time-point 'now', accessible for our perception, coexists with the volume of the related object or the system.

The Arrows of Time also display some specific characteristics associated with the type of the multimodal Space-Time and Info-Energy structure of a system. The direction of the Arrows of Time within TRMs of high energy massive radiating systems, such as stars, planets, star clusters and galaxies tangled by gravity (Figure 2), is different from the direction of the Arrows of Time within SRMs of spacious non-radiating systems (Figure 3).

In the TRMs of the radiating systems, the Space of the Progressive Time is a deviation from the Progressive Time with the characteristics of the time and time-associated energy 'flow'. The Arrow of the Progressive Time is directed from the '0' time-point at the Matrix centre to the system's future. The Space of the Regressive Time is a deviation from the Regressive Time with the characteristics of the time and time-associated energy 'contra-flow'. The Arrow of the Regressive Time is directed from the '0' time-point at the Matrix centre to the system's past (Figure 2).

In the SRMs of the non-radiating objects and systems, such as the thermodynamic Universe and the self-rising vacuum, the Space of the Progressive Time is a deviation from the Progressive Time with the characteristics of the time and time-associated energy 'contra-flow'. The Arrow of the Progressive Time is directed from the future of the system to the '0' time-point at the Matrix centre. The Space of the Regressive Time is a deviation from the Regressive Time with the characteristics of the time and time-associated energy 'flow'. The Arrow of the Regressive Time is directed from the past of the system to the '0' time-point at the Matrix centre (Figure 3).

The directions of the Arrows of Time provide us with an opportunity to detect possible changes of the Space-Time direction and reverse of the energy flow within the systems, for example, the Sun, our planet, systems holding black holes, and the Universe.

Matrix Balance

Isaac Newton defined inertia as his first law. He described the innate force of matter as 'a power of resisting by which every body, as much as in it lies, endeavours to preserve its present state...' [Newton Isaac, Mathematical Principles of Natural Philosophy (1687)]

Albert Einstein indicated, 'It was found that inertia is not a fundamental property of matter, nor, indeed, an irreducible magnitude, but a property of energy.' [Einstein Albert, A Brief Outline of the Development of the Theory of Relativity (1921)]

The modern understanding interprets inertia as 'a property of matter by which it continues in its existing state of rest or uniform motion in a straight line, unless that state is changed by an external force' (Oxford dictionary).

Inertia is a tendency to remain unchanged or resist to changes. It is, indeed, a property of energy. Every Matrix displays a tendency to resist any changes in its energy, associated information, and Space-Time. Every existing object and every system of the objects display a tendency to obtain and retain its Space-Time, Info-Energy, and Mass-Energy balance and symmetry about its Space-Time axis and 'o' Space-Time point at the centre of the Matrix.

Accordingly, the Matrix of an object or a system is symmetric about the Space-Time axis and 'o' Space-Time point at the Matrix centre if the Matrix is balanced.

The 'o' time-point, representing the time component of the 'o' Space-Time point, indicates the Matrix centre for us.

The 1-dimensional Space-Time and the associated Info-Energy are balanced and symmetric at the 'o' time-point at the centre of the Matrix if the Matrix is balanced. Accordingly, the Spaces of the Progressive and Regressive Time are symmetric at the 'o' time-point if the Matrix is balanced.

The Arrows of Time are directed along the Matrix axis, contra-directed, and symmetric at the 'o' time-point - the point 'now', if the Matrix is balanced that means the related multimodal object or the system is balanced in its Space-Time, Mass-Energy, and Info-Energy (Figures 2, 3).

The fundamental 2-dimensional Space-Time and associated Info-Energy of the 2-dimensional grid are balanced and symmetric about the Space-Time axis and 'o' time-point at the centre of the Matrix if the Matrix is balanced.

The human perception of time as a point 'now' provides us with the unlimited actuality of space and space-related objects and systems and makes available our interpretation of the object/system's volume. The system's Space-Time imbalance, relative to the 'o' Space-Time point at the centre of its Matrix, and associated Info-Energy and Mass-Energy imbalance provide us with the Space of the Current Time - the system's volume filled with mass, kinetic energy, and other energy representations acting in the volume of the system in the period of time now occurring.

In the balanced TRM of a radiating system, the Matrix resultant vector-force reflects the balanced influence of the Matrix grid vector-force of pressure developing the Matrix Spaces of the Progressive and Regressive Time (Figure 2). The typical TRM of a human body exists in a dynamic balance.

The balanced TRM of a radiating system displays the balanced tendency to transform the system's volume into the Matrix Spaces of the Progressive and Regressive Time, along with the tendency to transform the system's actual Info-Energy into the potential Info-Energy of the Matrix 2-dimensional grid forming the Spaces of the Progressive and Regressive Time. The TRM is balanced if its Space-Time and Info-Energy are balanced about the Space-Time axis and the centre of the Time-Rising Matrix.

In the balanced SRM of a non-radiating system, the Matrix resultant vector-force reflects the balanced influence of the system's vector-force of resistance developing the volume of the system in the Matrix Space of the Current Time (Figure 3). The SRM, existing in a dynamic balance, is the typical Matrix of the non-radiating, non-developing vacuum that does not significantly increase in size. The balanced SRM displays the balanced tendency to transform the Matrix Spaces of the Progressive and Regressive Time into the system's volume, along with a tendency to transform the time-associated potential Info-Energy of the Matrix grid into the system's actual Info-Energy. The SRM is balanced if its Space-Time and Info-Energy are balanced about the Space-Time axis and the centre of the Matrix.

The potential and actual Info-Energy structures counteract and keep a balance in the Matrix, following the Laws of Space-Time, Info-Energy, and Mass-Energy Conservation, Transformations, Reversibility, Limitation, and Balance and Symmetry, or Inertia as Newton's first law and the principle of parsimony, the scientific principle that states that things are usually connected or behave most simply or economically.

The Matrix resultant vector-force equals zero if the Matrix is balanced and symmetric about the Space-Time axis and 'o' Space-Time point.

The existing objects and systems are tangled together by the Space-Time, Info-Energy, and Mass-Energy imbalance.

Planck units, such as Planck length, Planck time, and Planck energy indicating the minimal Space-Time and Info-Energy imbalance, are to be understood as the minimal conditions of an object or a system's existence in our dynamic world. Planck length, Planck time, and Planck energy, and balancing forces, acting between dissimilar dimensional layers of the Matrix structure, are the factors influencing the directions of the Space-Time and Info-Energy 'flow' and the reverse of these directions in the existing multimodal objects and systems.

Image 9: 'Amazing Andromeda Galaxy' Image credit: NASA/JPL-Caltech

'The many "personalities" of our great galactic neighbour, the Andromeda galaxy, are exposed in this new composite image from NASA's Galaxy Evolution Explorer and the Spitzer Space Telescope... Located 2.5 million light-years away, the Andromeda is our largest nearby galactic neighbour.' NASA

Space-Time Imbalance

The Space-Time imbalance is often related to the specific Mass-Energy and associated with the geometry of the high energy massive radiating systems and large spacious non-radiating systems.

Every Matrix demonstrates a tendency to obtain and maintain a balance. Two opposite vector-forces act in the Space of the Current Time at the Matrix centre - the Matrix grid vector-force of pressure and the system's vector-force of resistance. The resultant vector-force of the unbalanced Matrix reflects either the dominant influence of the Matrix grid vector-force of pressure or the dominant influence of the system's vector-force of resistance. The influence of the Matrix forces brings about the balancing Space-Time, Info-Energy and Mass-Energy transformations.

The Space-Time imbalance, such as the excessive time and deficit of space, and the associated Mass-Energy and Info-Energy imbalance, humans perceive as gravity. The Matrix imbalance leads to the balancing Space-Time, Info-Energy and Mass-Energy transformations.

If we apply to the Matrix of the considered Space-Time region the point-of-time settings associated with our perception of time or the linear time settings in accordance with our understanding of time as linear time with the time 'flow' from the past to the future, there is no any sign of the gravitational field or antigravity that would generate accelerated or decelerated motion relative to an observer located outside of the Matrix or an observer located at the '0' time point at the Matrix centre.

In the second time dimension (Figures 4, 5), the centre of a system is directed to the Matrix Space of the Regressive Time while its surface is situated facing the Matrix Space of the Progressive Time.

The second time dimension reflects the function of the human consciousness on the past edge of the sensory input while developing a mental conception on the base of the sensory data and previously learned information.

Applying the second time dimension to the objects and systems in their Matrixes, we detect the Space-Time and Info-Energy imbalance in high energy massive radiating systems and spacious non-radiating systems. For example, the systems, holding black holes, demonstrate the Space-Time and Info-Energy imbalance in the second time dimension facing our dynamic world.

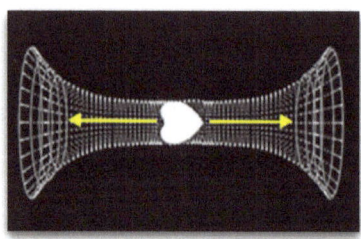

Figure 4: Time-Rising Matrix (TRM) imbalance

The Arrows of Time are not harmonically balanced in the Matrixes of these objects and systems in the second time dimension and 2-dimensional time settings.

The Matrixes display the curved line of the structured representation of the unbalanced function, reflected in the Space-Time and Info-Energy imbalance, gravity, antigravity, and geometry of unbalanced objects and systems.

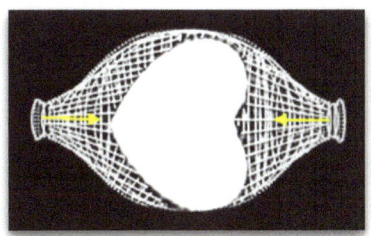

Figure 5: Space-Rising Matrix (SRM) imbalance

In the second time dimension and 2-dimensional time settings, the excessive time is directed from the surface of a high energy radiating system (Figure 4), such as the Earth, and the time deficit is directed from the centre. The excessive time accompanies the deficit of space and vice versa. In our dynamic world, the excessive time and potential space on the surface of the high energy massive radiating systems are supported by Info-Energy imbalance, such as excessive potential energy and the deficit of the mass, kinetic energy and other energy representations acting in the peripheral areas of the system, along with the excessive mass and deficit of the potential Info-Energy at the centre of high energy massive radiating system in the period of time now occurring.

In the second time dimension and 2-dimensional time settings, the excessive time is directed to the centres of the Space-Rising non-radiating systems (Figure 5), such as the Universe, and the time deficit is directed to their surface. The associated Info-Energy imbalance is reflected in the excessive mass and the deficit of potential energy in the peripheral areas of the Space-Rising non-radiating systems and the excessive time or potential space, supported by the potential Info-Energy of the system, along with the deficit of actual space and mass, kinetic energy and other energy representations at the centre of the system in our dynamic world.

Besides, applying the second time dimension to an unbalanced system in its Matrix, we detect a new centre of the Matrix symmetry - a new '0' time-point at the centre of the Matrix. Different representations of the same '0' time-point correspond to the realization of the same Matrix seen in two different time dimensions (the TRM in Figures 2, 4 and the SRM in Figures 3, 5). We can detect two different representations of the same '0' time-point, which coexist in the 2-dimensional time settings in the multimodal Space-Time of the unbalanced objects and systems.

The coexistence of two different representations of the '0' time-point in the TRM confirms a coexistence of the qualities of a toroid along with the qualities of a globe in the high energy massive radiating systems, such as our planet, in the multimodal Space-Time, and accordingly, the development of their geometry as the degeneration of a toroid in our dynamic world. The qualities of a toroid are related to the past of the unbalanced radiating system, and the qualities of a globe are associated with the system's future. The geometry of the system in the multimodal Space-Time is reflected in its specific geometry in our dynamic world.

The coexistence of two different representations of the '0' time-point in the SRM confirms a coexistence of the qualities of a toroid along with the qualities of a globe in the Space-Rising non-radiating systems, such the Universe, in the multimodal Space-Time, reflecting the evolution of these system's geometry as the developing toroid in our dynamic world in the future. The qualities of a globe in the past and the qualities of a toroid in the future are reflected in the specific geometry of the thermodynamic Universe.

Our understanding of the objects characteristics, motion, acceleration, and forces acting in our dynamic world, applies to the Space of the Current Time.

The system's Space of the Current Time includes the curvature of the system's Space-Time affected by gravity.

According to Albert Einstein, gravity is a consequence of the curvature of Space-Time caused by the uneven distribution of mass.

We mentioned above that the external boundary of the curvature of the system's Space-Time, affected by gravity, surrounds the total volume of the system. We refer the external boundary of the curvature of the system's Space-Time affected by gravity as 'the surface of the system' beyond which events, such as gravity or antigravity, cannot affect an observer.

Besides, the particular geometry of the system in our dynamic world reflects the system's specific toroid-globe geometry and the associated uneven distribution of mass.

Gravity and Antigravity

The specific Space-Time imbalance and the associated Mass-Energy, and Info-Energy imbalance are perceived as gravity. We observe and experience gravity on the surface of the Earth. We can detect signs of gravity around other high energy massive radiating objects and systems, such as the Sun, other stars, planets, globular star clusters and galaxies tangled by gravity.

The Space-Time, Info-Energy, and Mass-Energy imbalance, reflected in gravity, result in the forces performing balancing transformations reflected in gravitational acceleration.

Neither Mass-Energy nor the forces but the gravitational acceleration is the definite proof of gravity existence. The gravitational acceleration of the object's fall does not depend on the mass of the object. Mass and energy arrange the detectable physical basis for the Space-Time imbalance and balancing transformations.

However, the gravitational acceleration is the quality of the 'gravitating' system. In case of the Earth, the value of the gravitational acceleration is specific to the planet.

In the process of the gravitational acceleration, we can detect the unbalanced relationship between space and time as the linear space per time squared. On the Earth, objects fall with a standard value of acceleration 9.80665 m/s^2. Nevertheless, at different points on the Earth, objects fall with an acceleration between 9.764 m/s^2 and 9.834 m/s^2 depending on altitude and latitude. The gravitational acceleration increases from about 9.780 m/s² at the equator to about 9.832 m/s² at the poles, because the relations between space and time are various at different points on the surface of the Earth and the geometry of the planet is affected by its geometry in the 2-dimensional time settings.

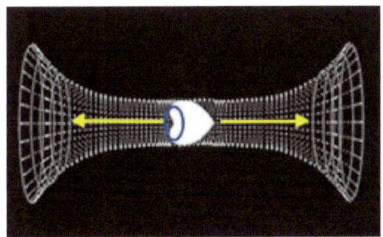

Figure 6: TRM Gravity

The surface of the high energy massive radiating system and the centre of the Space-Rising non-radiating system in the second time dimension are marked (Figure 6, 7). Please don't project the 3-dimensional understanding of space directly upon the system in the second time dimension.

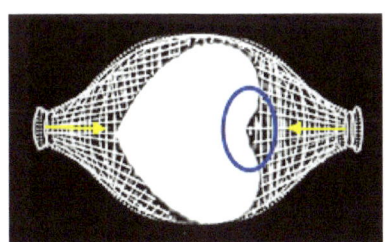

Figure 7: SRM Gravity

Gravitational acceleration, reflecting the relationship between space and time, is the element that let us recognise the gravity and the degree of gravity, measured by acceleration.

The primary process in gravity propagation is associated with the Space-Time transformations and the associated Info-Energy and Mass-Energy transformations.

The transformation of the excessive time and time-associated potential energy on the surface of the system into the volume, mass, kinetic energy and other energy representations, acting on the volume of the system in the period of time now occurring, are reflected in the gravity propagation in our dynamic world.

The process of gravity propagation is represented in the concentration and consolidation of the volume of the system, along with the concentration and consolidation of mass, kinetic energy, and other energy representations filling up the volume of the system in the period of time now occurring.

The 'gravitational horizon' is the 'transformations horizon' - the Matrix Info-Energy grid, which is arranged by the 2-dimensional representations of background radiation. It is the leading element in the Space-Time, Info-Energy, and Mass-Energy transformations. Matrix forces arise along the 2-dimensional Info-Energy grid forming Spaces of Time. The influence of the system's vector-force of resistance upon the Matrix Info-Energy grid concentrates and consolidates the volume and masses of our planet, retaining the total Mass-Energy and Info-Energy of the Matrix unchanged under the natural laws mentioned above.

Antigravity reflects the Space-Time imbalance, such as the excessive space and time deficit, and associated Info-Energy and Mass-Energy imbalance, such as the excessive mass and deficit of potential energy at the centres of the high energy massive radiating objects and systems, such as our planet, and in the peripheral areas of the Space-Rising non-radiating objects and systems, such as the Universe.

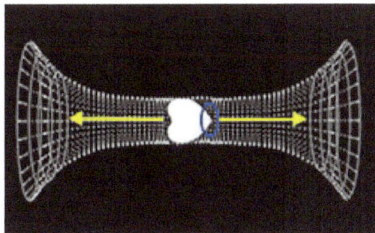

Figure 8: TRM Antigravity

The centre of the high energy massive radiating system and surface of the Space-Rising non-radiating system in the second time dimension are marked (Figure 8, 9).

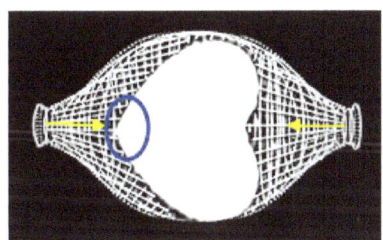

Figure 9: SRM Antigravity

Anti-gravitational deceleration, reflecting the relationship between space and time, lets us recognise the antigravity and the degree of antigravity, measured by acceleration.

The primary process in antigravity propagation is associated with the Space-Time transformations and the associated Info-Energy and Mass-Energy transformations.

Transformations of the excessive, super-concentrated space into the potential space and time, proceed along with the transformation of the excessive mass and energy into the potential energy associated with the potential space and time of the system.

The process of antigravity propagation is associated with the degeneration and reduction of volume, mass, and kinetic energy of the system, along rising of the system's potential resources.

The generation of space, energy, and matter within the system under the influence of gravity, coexists in a dynamic balance with the degeneration of space, energy, and matter into different forms of non-radiating degenerated matter and energy and their reduction prompted by antigravity.

The Equivalence principle

The Equivalence principle of General Relativity states that 'at any point of Space-Time the effects of a gravitational field cannot be experimentally distinguished from those due to an accelerated frame of reference' (Oxford dictionary).

We, supporting the Equivalence principle of General Relativity, view the mentioned gravitational field as the Space-Time imbalance, such as the excessive time and deficit of space supported by the associated Info-Energy and Mass-Energy imbalance. At any point of Space-Time, the effects of the specific Space-Time imbalance, such as the excessive time and deficit of space, supported by the Info-Energy and Mass-Energy imbalance and reflected in gravity, cannot be experimentally distinguished from those due to an accelerated frame of reference.

We, developing the Equivalence principle of General Relativity, view antigravity as the specific Space-Time imbalance, such as the excessive space and deficit of time supported by the Info-Energy and Mass-Energy imbalance. At any point of Space-Time, the effects of the specific Space-

Time imbalance, such as the excessive space and deficit of time, supported by the Info-Energy and Mass-Energy imbalance and reflected in antigravity, cannot be experimentally distinguished from those due to a decelerated frame of reference.

Image 10: 'Where Galactic Snakes Live' Image credit: NASA/JPL-Caltech 'Spitzer's new view of the region provides the best look yet at the massive embryonic stars hiding inside the snake. Astronomers say these observations will ultimately help them better understand how massive stars form. By studying the clustering and range of masses of the stellar embryos, they hope to determine if the stars were born in the same way that our low-mass sun was formed - out of a collapsing cloud of gas and dust - or by another mechanism... The snake is located about 11,000 light-years away in the constellation Sagittarius.' NASA

Balancing forces

The unbalanced systems, existing in the Universe, are not spherically symmetric. The specific geometry of the system is reflected, for example, the effects of the specific Space-Time imbalance and the associated Mass-Energy and Info-Energy imbalance. The Matrix imbalance exposes itself in gravitational acceleration and anti-gravitational deceleration.

The unbalanced multimodal Space-Time and Info-Energy structures display a tendency to obtain and retain a balance and symmetry through the '0' Space-Time point at the centre of the Matrix and the Space-Time axis.

Balancing transformations are initiated by the influence of the two opposite vector-forces acting in a dynamic balance at the Matrix centre - the Matrix grid vector-force of pressure and the system's vector-force of resistance.

Balancing forces emerge within the fundamental 2-dimensional Space-Time. The Matrix grid vector-force of pressure diverges from the system's vector-force of resistance within the 2-dimensional grid arranged by background radiation.

The Matrix grid vector-force of pressure limits the system's volume in the Matrix Space of the Current Time and develops the system's Spaces of the Progressive and Regressive Time.

The Matrix grid vector-force of pressure applies pressure (contact force) on the surface of the system enclosed at the centre of the Matrix. This vector-force has a direction perpendicular to the surface of contact (normal force). This pressure is a measure of the system's latent, or potential, Info-Energy stored in the Matrix grid. It is related to the latent energy density and latent information density associated with the system's potential space and time of existence.

The potential space, formed within the 2-dimensional grid and containing space-associated latent Info-Energy of the system, is a transmitter between potential and actual structures of the system. It functions similar to the restricted genetic code expression. The 2-dimensional grid keeps and transforms energy and associated information in compliance with the stored data.

Info-Energy blocks, which are processed, stored and transmitted by the 2-dimensional grid, alter the system's structures and functioning in a manner specific to the system.

The influence of the grid vector-force of pressure upon the system's volume is associated with the Matrix tendency to transform the Space of the Current Time, or the volume of the system, into the Matrix Spaces of the Progressive and Regressive Time, along with the tendency to transform the actual Info-Energy, acting in the volume of the system as mass and kinetic energy, into the potential Info-Energy of the 2-dimensional grid, retaining the total Space-Time, Mass-Energy, and Info-Energy of the Matrix unchanged following the laws of Space-Time, Info-Energy, and Mass-Energy Conservation, Transformations, Reversibility, Limitation, and Balance and Symmetry.

The system's vector-force of resistance is a measure of the system's actual Info-Energy acting in the volume of the system in the period of time now occurring. The system's actual Info-Energy, bounded with the Matrix grid in the Space of the Current Time, results in the system's vector-force of resistance that influences the 2-dimensional grid (contact force). This vector-force has a direction perpendicular to the surface of contact (normal force). The system's vector-force of resistance develops the Space of the Current Time.

The influence of the system's vector-force of resistance upon the 2-dimensional grid is associated with the Matrix tendency to transform the Spaces of the Progressive and Regressive Time into the Space of the Current Time, or the volume of the system, along with the tendency to transform the potential, Info-Energy of the 2-dimensional grid into the system's actual Info-Energy, retaining the total Space-Time, Mass-Energy, and Info-Energy of the Matrix unchanged.

The Space-Time imbalance, supported by the Mass-Energy and Info-Energy imbalance, specific geometry of the unbalanced multimodal systems, and influence of the Matrix forces, acting between dissimilar dimensional layers of the system, lead to the specific rotation of these systems about their Space-Time axis and the associated rotation in our dynamic world.

The Space-Time, Mass-Energy, and Info-Energy imbalance, reflected in antigravity, is responsible for the dominance of the 2-dimensional grid vector-force of pressure at the centres of high energy massive radiating systems and on the periphery of Space-Rising non-radiating system.

The Space-Time, Mass-Energy, and Info-Energy imbalance, reflected in gravity, is responsible for the dominance of the system's vector-force of resistance at the centres of Space-Rising non-radiating systems and on the periphery of high energy massive radiating systems.

The balancing flux of radiation diverges within the system's 2-dimensional Space-Time across the 2-dimensional Info-Energy grid arranged by the 2-dimensional representations of background radiation. Dynamic representations of background radiation arrange the internal framework and Space-Time and Info-Energy skeleton of the systems existing in our dynamic world. Dynamic representations of background radiation arrange, support, transport, and transmit balancing transformations, associated with gravity and antigravity.

Space, energy, and mass, generated from the potentiality under the influences of gravity, degenerate throughout the system into inflated space and different forms of non-radiating matter and energy and being reduced to potentiality under the influence of antigravity.

The presence of non-radiating matter and energy indicates the anti-gravitational processes in the area.

The difference between the directions of the time and time-associated energy flow within radiating and non-radiating systems leads to different qualities of matter and energy degenerated under the influence of antigravity within high energy massive radiating systems and Space-Rising non-radiating systems. We call the degenerated matter and energy, associated with anti-gravitational processes within the central regions of the high energy massive radiating systems, 'Black matter' and 'Black energy' to express the difference with 'Dark matter' and 'Dark energy'.

The difference in qualities of Black matter and Dark matter is comparable to the difference in qualities of the degenerated and dissociated matter at the centres of the high energy massive radiating systems and the vacuum of outer space. Probably, the elements of the degenerated matter and energy are not so far from us as we think.

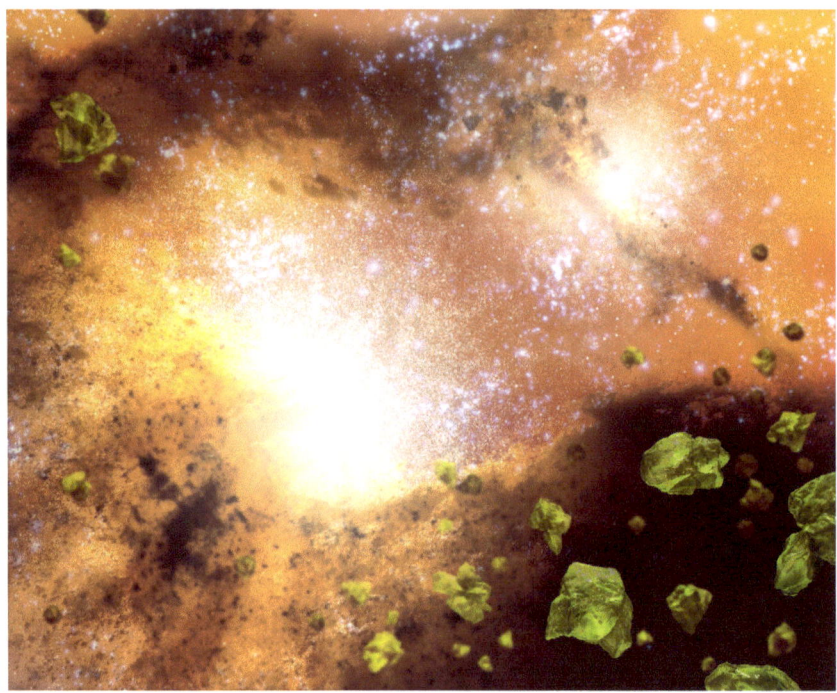

Image 11: Galactic Hearts of Glass (Artist Concept) Image credit: NASA/JPL-Caltech 'When galaxies collide, they trigger the birth of large numbers of massive stars. Astronomers believe these blazing hot stars act like furnaces to produce silicate crystals in the same way that glass is made from sand. The stars probably shed the crystals as they age, and as they blow apart in supernovae explosions. At the same time the crystals are being churned out, they are also being destroyed. Fast-moving particles from supernova blasts easily convert silicates crystals back to their amorphous, or shapeless, form.' NASA

The Main Principles

Principles of the Theory of Matrix apply to every existing object and every system of the objects, including visually 'empty' spaces, quanta, subatomic particles and holes, systems holding black holes and creating supernovae, other objects and systems existing in the Universe.

1. Neither empty space nor empty time exists in the Universe. The space of the Universe is formed by and filled with energy. We perceive this energy in different forms, such as masses, kinetic energy, and other energy representations acting in the volumes of the objects and systems existing in the Universe in the period of time now occurring. The potential energy and associated information build and fill the time of the Universe.

Following our perception of time, we cannot directly sense the time-associated energy of the Universe, but we can detect the influence of the potential energy as and when we experience gravity. The latent, or potential, Info-Energy is the predominant form of energy in the Universe. The latent, or potential, energy of the objects and systems, existing in the Universe, is coded and fixed by the latent, or potential, information within the different representations of background radiation.

Background radiation, arranging the potential Info-Energy into the 2-dimensional framework within the multimodal structures of the objects and systems, builds simultaneously the internal framework and Space-Time and Info-Energy skeleton for every object and system existing in our dynamic world. Background radiation and light constitute the background structure of our dynamic world.

2. Multimodal objects and systems of the multimodal Universe

Every existing single object and every system of the objects, including visually 'empty' spaces, subatomic particles and holes, quanta, humans, systems holding black holes, other objects and systems, and our Universe, are the multimodal objects and systems carrying the properties of various Space-Time modalities. The object or the system's characteristics are different in different modalities of Space-Time.

The multimodal Space-Time and Info-Energy structure of an object or a system we call the object or the system's Matrix for short. Accordingly, the Matrixes are the multimodal Space-Time and Info-Energy structures of the objects and systems with qualities of mass, volume, energy, and time of existence.

The 1-dimensional Space-Time, supported by energy, is an undifferentiated existence of space, time, and energy underlying our dynamic world. It is built by and filled with the one quality Info-Energy of infinite duration, presenting our dynamic world with the ultimate dynamics.

The object or the system's fundamental 2-dimensional Space-Time is built by and filled with the potential Info-Energy of the object or the system. The 2-dimensional time forms the object or the system's Spaces of Time. The 2-dimensional potential space underlies the volume of the object or the system.

The 3-dimensional Space-Time, or the volume of an object or a system, is built by and filled with the actual Info-Energy represented in mass, kinetic energy and other energy representations acting in the volume of the object or the system in the period of time now occurring.

3. Building blocks (STIE-blocks) of the Universe

The multimodal Space-Time and Info-Energy structures of the existing objects and systems are tangled together by the Space-Time, Info-Energy, and Mass-Energy imbalance and balancing transformations transmitted by background radiation. Dynamic representations of background radiation make the Matrixes the 'space-time-info-energy' building blocks (STIE-blocks) of the Universe.

The study of background radiation and its energy characteristics in the 2-dimensional Space-Time settings will provide a vast spectrum of opportunities for a broad range of the natural sciences.

4. Space-Time Equivalence

The universal proportionality exists between equivalent amounts of energy and mass,

$$E = mc^2,$$

where E is energy and m is mass.

The universal proportionality exists between equivalent amounts of time and space,

$$Et^2 = [c]^2 ml^2,$$

where E is energy, m is mass, l^2 is the 2-dimensional space, t^2 is the 2-dimensional time, and $[c]^2$ is the Coefficient of Transformation.

5. The Coefficient of Transformation $[c]^2$

The Coefficient of Transformation $[c]^2$ equals the squared numerical value of the speed of light in a vacuum.

The Coefficient of Transformation $[c]^2$ applies to the decisions associated with the Space-Time, Info-Energy, and Mass-Energy transformations. Dynamics preserve the Laws of Space-Time, Info-Energy, and Mass-Energy Conservation, Reversibility, Transformation, Limitation, and Symmetry.

The value of the speed of light in a vacuum (c) is 299 792 458 m s-1 (meter per second). It is the Fundamental Physical Constant. Under CODATA, the velocity of light measurements carried out by Michelson and his associates over the period 1924 to 1935. (NIST. Fundamental Physical Constants)

6. Relativity of motion

The concept of relativity and Einstein's Special Theory of Relativity state that all motion is relative and the velocity of light in a vacuum has a constant value which nothing can exceed.

We add that the relative difference between our perception of space and time makes this world dynamic for us - relations between space and time we perceive as speed and acceleration.

The speed of light (and background radiation) in a vacuum as the fundamental relation between 1-dimensional space and 1-dimensional time defines the exact proportion of space to time as the upper limit of the existing objects and systems and the Universe as we measure and sense it. Planck units, such as Planck length, Planck time, and Planck energy based on the calculations using the speed of light in a vacuum as the fundamentals proven by experiments, provide the lower limits of the existing objects and systems and the Universe as we measure and sense it.

7. The Space-Time Coefficient [c]

All dynamics in our dynamic world is associated with the difference in our perception of space and time - the relation between 1-dimensional space and 1-dimensional time we perceive as speed. The numerical value of the velocity of light in a vacuum is the Space-Time Coefficient that defines the relation between 1-dimensional space and 1-dimensional time in a vacuum.

The Space-Time Coefficient [c] equals the numerical value of the speed of electromagnetic waves propagation in a vacuum.

The Space-Time Coefficient [c] applies to the decisions associated with the transmission of the Space-Time, Mass-Energy, and Info-Energy transformations, including those reflected in gravity and antigravity propagation in our dynamic world.

8. Relativity of time, space, mass, energy, and information

Time, space, mass, energy, and information do not exist independent of objects and systems of the objects. They are the properties of the existing objects and systems. Time, space, mass, energy, and information are to be defined in relation to a frame of reference.

Space-Time is to be defined in relation to a frame of reference. Objects and systems of limited volumes are limited in their time of existence and, accordingly, limited in Space-Time. Objects and systems of the limited time of existence are limited in Space-Time. We specify the relation between time and space. Time is relative to space, changing, and transforming into space, and space is relative to time, changing, and transforming into time.

Info-Energy is to be defined in relation to a frame of reference. Mass-Energy is 'the mass of a body regarded as energy, according to the laws of relativity' (Oxford Dictionary). Objects and systems of limited mass are limited in their energy, and objects and systems of limited energy are limited in their mass. Energy is relative to masses, changing, and transforming into masses, and masses are relative to energy, changing, and transforming into energy.

Objects and systems of limited mass and energy are limited in Space-Time.

Objects and systems of limited volumes and time of existence are limited in their Info-Energy.

9. Relation of time, space, mass, energy, and information to the observer

The time, space, mass, energy, and information of the existing objects and systems are relative to the observer. You remember the Schrödinger's cat. Only the observer could say the cat in the box is either alive or dead. Alternatively, the cat is not in the box in the first place.

The observer observes an event and measures it, but the normal limits of human perception of the event and the environment restrict his ability to observe the existing object or the system. The observer's observation abilities are limited in quantity (5) and quality of senses. We, humans, observe the world and the existing objects and systems using our senses, bodies, and memories. We use the receptors of our bodies with their limited ability to respond to the external stimuli, and it takes time to transmit the signal by the series of synapses, compare it with the preserved data in several layers of memories, and produce the sensible response.

We reproduce our limited diapason of senses in new technologies extending the quality of our perceptual abilities. Nevertheless, new technologies reflect our limited quantity of senses.

Accordingly, the world is relative to the observer and not defined.

10. The Systems Theory and the Principle of Balance in Cosmology

Space, time, mass, energy, and information of every object are integrated into space, time, mass, energy, and information of the system, similar to the water molecules in the ocean or points of space included in the set of points having some specified structure in mathematics.

Space, time, mass, energy, and information of an object in the system is entirely negligible relative to space, time, mass, energy, and information of the system and reflected in the resultant space, time, mass, energy, and information of the system. Similarly, water molecules are bonded together in the ocean, thus achieving a new quality.

Forces between objects in a system are integrated and reflected in the resultant force of the system that is built by these objects. Forces between objects in the system are entirely negligible relative to the resultant force of the system built by these objects.

Space, time, mass, energy, and information of the existing objects and systems are integrated and reflected in the resultant space, time, mass, energy, and information of the Universe. Forces between the existing objects and systems are integrated and reflected in the resultant force of the thermodynamic Universe.

The Space-Time, Info-Energy, and forces of the multiple Universe modalities, such as the thermodynamic Universe and other Universes, Multiverse, stages of the Universe development, and other modalities of the Grand Universe, are finally integrated into the Space-Time, Info-Energy, and force of the Grand Universe.

The Grand Universe includes and balances all modalities of the Universe, and its 'o' Space-Time point, as the point 'here' and 'now', is the theoretical space-time-mass-energy null-point. It is the 'o' world point, or 'a space-time-null-point', mentioned in Special Relativity by H. Minkowski, A. Einstein, and H. Weyl in relation to the mathematical model of the Light Cone.

Actualities of the various existing modalities of the Universe, their potentialities, other properties, and numerous opposing forces are balanced about the theoretical 'o' space-time-mass-energy point and reflected in the resultant force of the Grand Universe. The resultant force of the Grand Universe equals zero.

The characteristics of different modalities of the Universe and various stages of the Universe development in complex combinations co-exist in the real multimodal world in the state of equilibrium, yielding a net-zero and providing us with the theoretical 'o' space-time-info-energy point - the world point. The 'o' point is the point of the absolute balance - the total equilibrium that can be applied to the Grand Universe and fails our dynamic world.

The multimodal Universe exists in the state of equilibrium that means the qualities cancel out, yielding a net quality of zero, thus making the Grand Universe neutral.

Accordingly, the multimodal Universe was not born. It has no start of the existence in space, time, mass, energy, and information. It has no end but modalities. This theoretically possible scenario could be applied only to the Great Multimodal Universe. A balanced, isolated object or an isolated system in the state of equilibrium would cease to exist.

We will reappraise these principles in the following book as different cosmological particulars exemplify them.

Prospects

The Theory of Matrix applies to every object and every system of the objects, including visually 'empty' spaces, subatomic particles and holes, quanta, humans, black holes, other objects and systems existing in the Universe. The Theory of Matrix applies to the Universe and its Matrix.

The application of the Theory of Matrix provides an opportunity to access the additional Space-Time characteristics and properties of the multimodal objects and different regions of our Universe, their Space-Time, Mass-Energy and Info-Energy structures; Space-Time, Mass-Energy and Info-Energy transformations; characteristics of gravity, antigravity and specific rotation of cosmic background radiation.

Energies of background radiation, including those known as electromagnetic radiation, cosmic background radiation, such as the cosmic microwave background radiation, light, et cetera, and associated information, supporting the info-energetic grid of the multimodal Matrixes, make Matrixes the 'space-time-info-energy' building blocks, or STIE-blocks of the Universe.

The dynamic representations of background radiation are accessible for our investigation. The studies of background radiation and its energy characteristics in the 2-dimensional Space-Time settings allow to gain additional knowledge, associated with the structure of the Universe and objects existing in the Universe, and provide a vast spectrum of opportunities for a broad range of the natural sciences.

'I'm of the opinion, that physics is conceptual, not illustrative.' (Albert Einstein, The Bad Nauheim Debate (1920)

Conceptual physics (plural noun, treated as sing.) is the study and description of properties and interactions of space, time, matter, and energy in the conceptual non-mathematical form with emphasis on logical reasoning in order to derive fundamental laws of nature and obtain conclusions from these laws by a sequence of logical steps. It is a branch of theoretical physics.

The human momentary perception of time is reflected in the mathematics and other abstract sciences associated with numbers, quantities and space. It is applied to other disciplines, such as physics. Mathematics based on single numbers, where 1^3 still equals 1, will not help us overcome the barrier of pointed consciousness. Development of 3-dimensional mathematics is necessary.

The Theory of Matrix supports the next step of the Space-Time Theory development with respect to the 3-dimensional time, its practical application to the 3-dimensional space to merge into the new quality of 3-dimensional Space-Time understanding.

Afterword

The Theory of Matrix is a new theory combining the elements of psychology, cosmology, and astrophysics.

Dr Randles developed the program for psychological investigations of the Universal Matrix along with the development of the Coresynthesis Psychological Model in early 1990s.

The Theory of Matrix was introduced by Dr Audrey E. Randles in her work 'The Theory of Matrix' in 2012. When Dr Randles published her work, the Light Cone was still considered a specific case applicable to the flash of light. She brought forward the idea of the universality of the Space-Time structure of the objects and a new vision on Space-Time physics.

Following the analysis of the parallels and variations between the Universal Matrix and Light Cone, Dr Randles calls the Light Cone 'the Matrix of Light' and presents the Theory of Matrix as the relative importance for the true understanding of the multimodal world.

The Theory of Matrix introduces the new vision of the multidimensional world with respect to multidimensional time. Time is no longer seen as another dimension of space, nor as a momentary feature of an event but as a multidimensional element in its own right. Dr Randles associates space with actuality and time with potentiality. Therefore, the objects are viewed as the multimodal, multidimensional objects in Space-Time.

Books on Cosmology by Audrey E. Randles:
'Systems Theory in Cosmology' (2020)
'The Multimodal World' (2020)

'Black Holes and Supernovas' (2016)
'Grand Universe' (2016)
'Antigravity' (2015)
'Supernovas' (2015)
'The Primary Black Hole of the Universe' (2015)
'Energy in Cosmology' (2014)
'Gravity and Antigravity to the Point' (2014)
The Theory of Matrix series of books (2012 - 2013) includes the following books:
'Blocks of the Universe'
'Space and Time'
'Energy of Existence'
'Gravity and Rotation'
'Black Holes'
'Matrix of the Universe'
The new 2020 books on Cosmology include Kindle ebook and a paperback of the same title at Amazon's Book Store.

Content Use Policy
Content may be used for any purpose without prior permission, subject to the special cases noted below.
By downloading the material, the user agrees:
1. to use a credit line in connection with the content. Unless otherwise noted in the caption information for any content and images the credit line should be 'Audrey E. Randles, "The Theory of Matrix. Space and Time" (2013) red 2020'.
2. that we do not represent others who may claim to be authors or owners of copyright of any of the content, and make no warranties as to the quality of the content;
3. that we shall not be responsible for any loss or expenses resulting from the use of the content, and you release and hold us harmless from all liability arising from such use.
Special Cases:
This content is available for educational, journalistic, personal uses and scientific research following a scientific code of ethics.
Restrictions are placed on commercial uses. To obtain permission for commercial use, contact the copyright owner Dr Audrey E. Randles.

www.ingramcontent.com/pod-product-compliance
Lightning Source LLC
Chambersburg PA
CBHW040223220526
45473CB00001B/91